Amp 125

Master Handbook of 1001 Practical Electronic Circuits

Other TAB books by Kendall Webster Sessions, Jr.

Book No. 590 Practical Solid-State Principles & Projects

Book No. 613 New IC FET Principles & Projects

Book No. 621 The 2-Meter FM Repeater Circuits Handbook

Book No. 647 Stereo/Quad Hi-Fi Principles & Projects

Book No. 667 Miniature Projects for Electronic Hobbyists

Book No. 673 How to Be A Ham—Including Latest FCC Rules

Book No. 722 Amateur FM Conversion & Construction Projects

Book No. 756 Four-Channel Stereo—From Source to Sound—2nd Edition

Book No. 827 Amateur Radio Advanced Class License Study Guide

Book No. 851 Amateur Radio General Class License Study Guide

Book No. 873 Amateur Radio Novice Class License Study Guide